Table of Contents

Tip 1:	Calling All Parents
Tip 2:	This Is Not a One-Sided Journey
Tip 3:	Welcome! The Change Is Here
Tip 4:	You're Stressing Them Out
Tip 5:	Break the Ice
Tip 6:	Teachers Are Professionals Too
Tip 7:	The Big "F" Word
Tip 8:	Not My Child
Tip 9:	The Wonderful World of Expectations
Tip 10:	Respect Has Loss Its Ranking
Tip 11:	We Should All Be the Good Cops
Tip 12:	I Don't Have Time
Tip 13:	The Parent-Teacher Conference Effect
Tip 14:	Watch Your Tone
Tip 15:	No Exclusions on Inclusion
Tip 16:	You Need to Go
Tip 17:	Set a Date and Create Routines
Tip 18:	Is Your Child Prepared
Tip 19:	There Will Be Consequences
Tip 20:	Bullying Is Not Tolerated

"Unlocking Success Together: Empowering Parents with Expert Tips from Teachers!"

Dedication: This book is dedicated to my husband and children. Thank you for helping me make my journey to become a better educator, mommy, and me a good one. You inspire me to be the best. Everything I do is for you! To all of my teachers, I got you, and we got this! To all of my parents who cry tears of frustration and who are struggling with their children and don't know where to turn for support, I got you, too.

To all the teachers and parents who tirelessly support their scholar's growth and learning, may this book serve as a bridge to strengthen our partnership in nurturing the next generation.

Dear Parents:

Welcome to the "undiscovered truths" of what's going on in our schools. This guide is crafted from the heart of a teacher to parents and other teachers. As an educator, I have had the privilege of witnessing the profound impact of collaboration between teachers and parents in shaping the lives of children. In these pages, I offer thought-provoking insights and truths inside the classroom, reflections, and practical strategies to foster a strong partnership between home and school.

Tip 1: Calling All Parents

Parents, we need you! As children matriculate throughout their educational journey, parental support decreases. Some parents feel their children are old enough to communicate effectively with their teachers and independently manage their school responsibilities. But that is far from the truth. As children get older, they need more parental support and guidance.

The parent-teacher partnership is crucial for fostering a supportive, nurturing, and enriching environment where children can thrive academically, socially, and emotionally. It is a collaborative effort that recognizes the importance of working together to support the growth and development of every child. So here are some ways to partner with your child's teacher:

Attend Parent-Teacher Conferences:
Regularly attend parent-teacher conferences to discuss your child's progress, strengths, areas for improvement, and goals. These meetings provide valuable opportunities to exchange information, ask questions, and collaboratively develop strategies to support your child's learning.

Communicate Openly and Proactively:
Maintain open lines of communication with your child's teacher throughout the school year. Share any concerns, observations, or insights about your child's academic performance, social interactions, or emotional well-being. Likewise, stay receptive to feedback and suggestions from the teacher, especially when it comes to behavior. Children

don't want to get in trouble at home so they might tell a little white lie or a full-blown lie. They often have their interpretations of what was said or how something was communicated, which brings in chaos.

Participate in School Events and Activities:
Take an active role in school events, workshops, and activities organized by the school or parent-teacher association. These events provide opportunities to connect with other parents, build relationships with teachers and staff, and contribute to the school community. Sadly, some parents don't take the time to do this. More parents participate in their children's activities outside of school (dance, gymnastics, football, soccer, etc.) than in-school activities. Your children must understand that school is the priority and everything else is secondary. When priorities are not in order or enforced, students are less likely to give their 100%.

Volunteer in the Classroom or School:

Teachers might roll their eyes at me or think I'm crazy (I'm already laughing at myself). I know some parents can be nuances or overbearing, and we want to keep some of them far away. But parents are beneficial. Parents should offer their time and skills to volunteer in their child's classroom, library, or extracurricular activities. Involvement can include assisting with classroom projects, organizing field trips, or participating in reading groups. Parents, your presence demonstrates your commitment to supporting your child's education and strengthens your relationship with the teacher. If you are that parent who "does too much," "does the most," or is "extra," please stay home, or we'll call you!

Support Learning at Home:
Reinforce learning at home by establishing routines for homework, reading, and study time. Stay informed about the curriculum and academic expectations, and provide encouragement and assistance as needed. Create a conducive learning environment at home that encourages curiosity, exploration, and critical thinking. This environment can become more challenging as children reach high school, but don't throw in the towel. Find ways to introduce your child to new ways of learning. Remember, Gen Zers and Alphas are our pandemic babies, and they are unique. Their education was suddenly disrupted, and they had to adapt to a new way of thinking and doing. So, just like teachers and administrators had to scramble to figure out what works upon our return from the pandemic, you, the parent, must figure out what works at home.
Remember, to change with time. What worked five years ago might not work now.

Engage in Parent Education Opportunities:
Parents should also find learning opportunities like they do for their children. Take advantage of parent education workshops, seminars, and resources the school or community organizations provide. These opportunities cover parenting strategies, academic support, social-emotional learning, and technology use. Enhancing your knowledge and skills can better support your child's educational journey. The more you know, the more you and your child grow!

Participate in Decision-Making Processes: Stay informed about school policies, initiatives, and decision-making processes. Consider joining parent advisory committees, school councils, or parent-teacher associations to have a voice in shaping school policies, programs, and priorities. Your input and perspective are valuable contributions to the school community.

Celebrate Achievements and Milestones: Children love to be celebrated, primarily when they've worked hard and there is so much competition. Celebrate your child's academic achievements, progress, and milestones with their teacher. Acknowledge their efforts, growth, and perseverance in overcoming challenges. Celebrating successes reinforces positive behavior and motivates continued effort and engagement in learning. It also allows the child to develop practices and gain a sense of pride that can extend into generations.

Seek Additional Support as Needed: Pay attention to the red flags. If your child is experiencing academic, social, or emotional challenges, don't hesitate to contact the teacher or counselor for guidance and support. Together, you can explore strategies, resources, and interventions to address your child's needs and promote their well-being. We need you to show up and be a part of this team.

▌ *Parents, teachers want you to know that you are valuable in this partnership, and we need you.*

Tip 2: This is Not a One Sided Journey

Even though teachers wear many hats and serve in many roles, your child's teacher is your child's teacher, not the parent.

I've heard some parents say, "It's the teacher that's supposed to do x, y, and z." "You are the teacher, and I'm at work." "Why are you calling me? What do you want me to do?" These are not the responses that teachers want to hear when reaching out to parents about their students. However, the reactions are more common today than ever before. That is, if you even get a response.

Over the years, there has been a shift in teacher responsibilities and expectations.
Parents and teachers are responsible for children's foundation, nurturing, growth, learning, and development; both play pivotal roles in a child's development. While teachers provide formal education within the classroom, parents offer invaluable support, encouragement, and guidance at home.

Understand that education is not confined to the walls of a school but extends into the fabric of daily life. Learning moments emerge in everyday experiences, conversations, and interactions.
Each stage of childhood and adolescence presents unique opportunities and challenges. From the formative years of early childhood to the transformative period of adolescence, children undergo profound physical, cognitive, and emotional growth. Depending on your child's stage, their overly dramatic emotions can heavily stand out more to the point of frustration for all who have

to encounter them. And if their theatrics are weighing in on you at home, just imagine what the teachers are experiencing at school.

Each developmental stage is different and should be embraced and celebrated.

By embracing the shared responsibility of guiding children's educational journey and understanding the nuances of developmental stages, parents and teachers lay a solid foundation for collaboration, empathy, and partnership. Together, we create a nurturing ecosystem where children thrive, learn, and flourish to their fullest potential.

Teachers are looking for the parents to step up and step in. When I say this is not a one-sided journey, it's not. Everything cannot fall upon the teacher. It takes the parent and the teacher to join forces and make a unified stance on developing a solid partnership.

Tip 4: You're Stressing Them Out

Parents, you are stressing your kids the heck out! And that's saying it mildly. Not only do you stress the kids out, but you can stress the teachers out, too!

If you are a teacher, you may have heard some complaints that students have of their parents, like:
"My parents don't understand."
"They won't listen to me."
"I try talking to them, but they keep yelling and fussing at me."
"My parents want me to make all A's."

Kids just want to be kids, and teachers just want to teach. Can that genuinely happen without probable cause? Remember, kids just want to be kids. They want to enjoy their friends and hobbies, eat junk food, play video games, or be on their phones. But, with all of the extracurricular activities, pressure to earn all A's, and taking care of younger siblings, students are an emotional wreck.

For middle schoolers, kids are still trying to figure things out. The transition is a major one. Many want to return to elementary school, where they've been nurtured and handheld. Adapting to middle school can be uncomfortable, and children can resent or resist being in their new state.

Parents, please hear me when I say, "Take a step back!" That does not mean you don't care, but you should give your kids some breathing room. Check-in with your child's teacher to see what they are observing. When I see my students underperforming, I intentionally check in with them. That's where I learned how much pressure they have, and most of it stems from their parents, which spirals them into an emotional wreck.

If you don't believe me, here are some other comments that teachers hear from students:
"My parents will start yelling at me if I bring a B or a C home."
"I'm stressed out."
"I have to get into Harvard."
"I have so much homework, and I don't know how to keep up."
"My mom (step-mom) needs me to…"
"My dad (step-dad) said if I don't…, then…"
"My parents are expecting me to…"
"I have to take care of my little sisters when I get home, so I won't have time to…"
"I have practice after school and get home late, so I can't…"

Teachers get stressed out for the kids when hearing all of the unrealistic expectations that are being placed on them. Kids just want to be kids, not trophy pieces. Let your children know that it is ok if they don't earn all A's as long as they do their best. Tell them you are proud of them for the accomplishments and transitions they've endured. Give them a hug when they bring home a "C" or even a "D," especially if it's their best. If your child earns a failing grade, support them as they work to improve their grade. This is the best gift you can give your child.

Students will not master everything or every skill, and the progressive struggle is necessary. Allow the space to analyze where your child is and what they need to do to improve. Remember, children are growing and developing, so help them to problem-solve while teaching them to advocate for themselves and to ask for help.

Whew...the pressure of it all! Set realistic expectations, identify the source of the stress, and encourage open communication. Your child wants to open up to you, share their concerns, and feel like they are being heard. Help your children see that it's ok if they are not perfect at everything because perfection can bring needless stress.

🚪 *We want our students to be happy and prosperous, but teachers often see how parents can impede the growth and success of their children.*

Tip 5: Break the Ice

Many teachers want to develop a good relationship with parents and are very approachable. The initial step can be intimidating, depending on experience, and parties can bring unnecessary baggage into the new-teacher relationship.

You want to start on a good foot and allow the relationship to grow.

Breaking the ice with your child's teacher is essential in establishing a positive and collaborative relationship. When expressing issues or concerns about your child, it's more productive to approach the concerns politely. So don't be aggressive; instead:

Introduce Yourself: Start by introducing yourself to the teacher during a parent-teacher conference, back-to-school night, or via email. Share your name, your child's name, and any relevant information about your family or your child's interests.

Express Interest and Appreciation: Express genuine interest in your child's education and appreciate the teacher's role in supporting their learning. Let the teacher know that you value their expertise and dedication to teaching.

Share Information About Your Child: This is where you need to be honest. Provide the teacher with relevant information about your child, including their interests, strengths, areas for improvement, and any learning or behavioral challenges or preferences they may have. This information can help the teacher better understand and support your child's needs.

Offer Support and Collaboration: Let the teacher know you are committed to supporting your child's education and collaborating with them as a partner in the learning process. Ask how you can help the teacher inside and outside the classroom.

Be Open to Feedback and Suggestions: Be open to receiving feedback and suggestions from the teacher about your child's progress, behavior, and academic performance—approach feedback with a positive attitude and a willingness to work together to address any concerns or challenges.

Express Gratitude: Show appreciation for the teacher's time, effort, and dedication to your child's education. A simple thank you can go a long way in building a positive and supportive relationship.

🚪 *Humility is the prescription of the day, and it goes a long way. It's never too late to establish a healthy parent-teacher relationship. So take action now!*

Tip 6: Teachers Are Professionals Too

That's right! Teachers are professionals, too, but they have become some of the least respected professionals somewhere along the way.

It amazes me how some parents communicate with their children's teachers, trying to boss them around, sending threatening emails, and teaching their kids how to disrespect their teachers verbally and physically. Likewise, the students can be equally disrespectful when communicating with their teachers. Unfortunately, some parents and students deem these actions acceptable.

Consider this- Is it ok for employees to lay hands on their boss? Curse them out? To tell the boss what they're not going to do? Let's take it a step further. Is it okay for someone to email or call your supervisor without allowing you to correct an issue or resolve a matter? You get the picture. I'm sure that we all agree that the answer is "no." Did you know that students curse out, fight, talk back to, and do all types of other disrespectful things to their teachers? I'm sure you agree that this is unacceptable if done to parents, grandparents, coaches, or spiritual leaders. Contrary to some beliefs, this and so much more goes on in this profession, and some parents are totally on board and accepting of the behaviors of other parents and children. How disturbing!

In my early teaching days, I remember teaching in an elementary school in an underserved community where most of the parents were unemployed and had substance abuse issues. I'll never forget when a parent pulled their children out of one class to fight a teacher. Yes, this parent went into two different classrooms to pull her children out to fight a teacher. I could not believe my eyes as the commotion was taking place. If that's not enough, multiple parents threatened to physically harm teachers for weeks at that same school. Yes, this is sad and true, but it was the reality of that school's culture. Imagine someone approaching you at your job because they did not like a performance appraisal, weren't awarded a contract, or were opposed to your decision. A person's workplace should be safe for an employee, just as school should be for students. Thankfully, nothing transpired when these teachers were threatened, but they should be able to protect themselves.

Parents should acknowledge teachers as the professionals they are because they possess specialized knowledge, expertise, experience, and commitment to supporting student learning and development. Many who have earned a master's, doctoral, or Ph.D. are experts in special education, equity and inclusion, curriculum and development, instructional design, educational leadership, e-learning, reading and math, administration, instructional design, and many other niches that the average person does not possess. Advanced degrees, such as a Master's or Ph.D., deepen teachers' understanding of their subject matter, teaching methodologies, and educational theories. This more profound knowledge equips teachers with the necessary tools to deliver more comprehensive and effective instruction. Earning these degrees is no walk in the park and should be held to the same level as earning an MBA.

Teachers spend more hours in a day and week with their students; therefore, we see things that parents do not.

By recognizing and respecting teachers' professionalism, parents can cultivate strong partnerships that enhance student success, promote positive school environments, and contribute to the overall well-being of children and families.

🚪 *Teachers are professionals and experts and should be treated as such. Respect for the profession is long overdue. Teachers should be able to protect themselves when they are physically attacked by anyone who plans to do bodily harm but are strongly warned not to. Otherwise, they will leave in handcuffs and lose their jobs. It's such a shame! We'll touch more on respect in a later chapter.*

Tip 7: The Big "F"

Children have to go through the processes of life. There will be some ups and some downs.

Frederick Douglass once said, "If there is no struggle, there is no progress." Allow your child to navigate through the process, which sometimes means allowing them to fail. Oh my, there's that nasty "f" word. Failing is a necessary step to achieving success. Now, don't get me wrong. I'm not saying allow your child to earn straight failing grades or fail at something and be paralyzed. However, if failing grades are earned or a failed opportunity results from poor choices, they must learn from it. Trust me, they will be ok. I, too, had to endure my failures in school, which was the best thing for me. It taught me about me. I only knew how much confidence and determination I had once I failed. Parents can make it difficult for their children to grow when they jump in and run to the rescue. It's not helping the child at all.

Failure, while often perceived negatively, can be an influential teacher who imparts valuable lessons and fosters personal growth and resilience. Here are some positive lessons that can be gleaned from failing:

Resilience: Failure teaches resilience by showing students that setbacks are natural and can be overcome with perseverance and determination. Through facing and overcoming failures, individuals develop the strength and resilience to bounce back from adversity and continue pursuing their goals.

Self-Reflection: Failing encourages self-reflection and

introspection as students analyze what went wrong, identify areas for improvement, and learn from their mistakes. This process of self-examination fosters self-awareness, personal growth, and a deeper understanding of one's strengths and weaknesses.

Adaptability: Failure teaches adaptability by challenging students to adjust their approach, strategies, and mindset in response to changing circumstances or unexpected obstacles. It encourages flexibility and creativity in problem-solving and cultivates the ability to navigate uncertainty and ambiguity with resilience and confidence.

Humility: Experiencing failure fosters humility by reminding students of their limitations, fallibility, and the importance of learning from others. It humbles individuals by challenging their ego and encouraging them to seek feedback, guidance, and support from others to improve and grow.

Empathy: Failure cultivates empathy by fostering an understanding and appreciation of the struggles and challenges faced by others. It enables students to empathize with the experiences of others, offer support and encouragement, and build meaningful connections based on shared experiences of adversity and resilience.

Courage: Failure encourages students to take risks, embrace uncertainty, and step outside their comfort zones to pursue their goals and aspirations. It inspires individuals to face their fears, confront challenges head-on, and persevere in the face of obstacles with courage and determination.

Success Redefined: Failure challenges conventional

High schoolers finally come into their own by their junior year and understand the importance of finishing strong. Peer influences shape their attitudes, behavior, and decision-making. Not only do they explore their identity, but they also explore the world around them more. This group's identity is critical to their success and well-being. Physical changes and sexual characteristics are prevalent throughout this transformative period. High school is a dynamic and transformative period characterized by physical, cognitive, social, and emotional changes. It is also a time of exploration, growth, and self-discovery as teenagers navigate the complexities of adolescence and prepare for adulthood.

The transition from elementary to middle school to high school is a period of exploration, growth, and self-discovery. I always encourage my parents to 'respect the process,' whatever it may be.

Tip 3: Welcome! The Change Is Here

Yes! Yes! Yes! The change is here, so put on your seatbelt and get ready for the ride.

When children get to school, they change into someone they don't even know. Acceptance is essential, and no matter what you say or the consequences, being someone different outside of the home is always more appealing.

The elementary schooler is curious and explorative regarding their interests and interactions with others. They explore their strengths, weaknesses, and personal preferences, developing a sense of self-esteem and self-confidence. They also begin to understand their place within their families, peer groups, and larger communities. However, this group often needs help with the initial transition, expectations, and responsibilities of middle school. They still want the benefits of being the "babies" but not the leadership.

Middle schoolers experience the most significant change. Like the middle child, they don't know where to fit in, and those dreaded hormones kick into high gear. They do not know who they are, nor do you. You might start getting calls and emails from teachers, grades fall, and the once-quiet, introverted soul becomes more vocal and likes to play "tag" with someone of interest. I call this my wild bunch because the hallway is filled with risky danger zones and often filled with "wild" behavior, activities, and characteristics.

notions of success and encourages individuals to redefine success on their terms. It encourages students to embrace a growth mindset, focus on continuous improvement, and measure success not solely by outcomes but by the effort, learning, and resilience demonstrated in the face of failure.

Celebration of Effort: Failure celebrates effort, perseverance, and resilience as students demonstrate the courage to try, the determination to persist, and the resilience to learn and grow from setbacks. It shifts the focus from the fear of failure to the value of effort and resilience in pursuing meaningful goals and aspirations.

On the brighter side, failure is not the opposite of success but an integral part of the journey toward personal growth, resilience, and fulfillment. Learn to embrace it and not try to manipulate or bully the teacher.

Contrary to some parents' beliefs, many teachers want to see their students do well and are not out to get them. If your child is failing or is struggling in a particular area, don't always look at the teacher. Look at your child and hold them accountable. The amount of time spent emailing the teacher can be used to converse with your child.

Tip 8: Not My Child

Yes, your child! Believe me when I say that children have their alter egos in school.

As soon as they get out of the car, off that bus, and into those four walls, they become different kinds of kids. Parents are surprised or act surprised and find it hard to believe certain things their children involve themselves with at school. Some behaviors are pleasing, while others are disappointing. If you're unsure where your child fits in, here are a few examples:

Misbehavior: Parents might be shocked to learn that their child misbehaved or received disciplinary action at school. Some behaviors could include talking back to the teacher, disrupting class, or engaging in inappropriate behavior with peers.

Academic Struggles: Parents might find themselves perplexed when they realize their child faces academic challenges, mainly if they believed their child was excelling. Discovering that their child is lagging in specific subjects or failing to complete assignments can be worrying and catch parents off guard.

Social Challenges: No parent wants to learn that their child faces social challenges at school, such as difficulty making friends, experiencing bullying, or feeling excluded from social activities. It can be distressing for parents to realize their child is having difficulty fitting in or navigating social interactions with peers.

Skipping Classes: Parents may be stunned that their child has been skipping classes or cutting school without their knowledge. Discovering that their child is not attending classes can be alarming and may indicate underlying issues that must be addressed.

Involvement in Negative Peer Groups: When a child is associating with pessimistic peer groups or engaging in risky behaviors, such as experimenting with drugs and alcohol or engaging in delinquent activities, it's gutwrenching. Parents must accept that their child is making poor choices influenced by peer pressure.

Dishonesty or Cheating: It's disappointing to discover when a child has been dishonest or engaged in cheating behaviors, such as plagiarizing assignments, copying answers during exams, or lying about school-related matters. Learning that their child has compromised their integrity can be distressing for parents.

Inappropriate Language or Content: It can be unsettling for parents to discover that their child is exposed to mature or inappropriate content. Parents may be surprised to hear that their child has been using unacceptable language or accessing inappropriate content at school, whether through conversations with peers or online activities.

Unusual Talents or Interests: On a more positive note, parents may be pleasantly surprised to learn about their child's unique talents, interests, or achievements at school. These talents could include discovering a hidden talent for art, music, or sports or excelling in academic subjects that they had yet to recognize previously.

Teachers see and hear things that are disturbing all day, every day, ranging from gang fights to drugs to things that are sexual. When teachers share what they are witnessing, it is a good time for parents to act.

Tip 9: The Wonderful World of Expectations

Parents and teachers should have high expectations of their students, which helps to catapult them into more significant areas of their lives. But that requires teaching accountability, responsibility, respect, and good work habits.

High expectations from teachers can be incredibly beneficial for students' academic and personal development. Here's why: It promotes growth, builds confidence, encourages persistence, prepares them for real-world challenges, fosters a growth mindset, reduces achievement gaps, inspires lifelong learning, and, most importantly, creates a culture of excellence.

When I think about high expectations, I don't equate them to perfection; rather, excellence. Excellence is a distinction, high quality, superiority, brilliance, greatness, merit, mastery, accomplishment, and proficiency. To put it plainly, excellence has outstanding features and quality. So how can students be the best in schools, top athletes, be gifted and talented, and have exceptional skills and hobbies if the expectation isn't one of excellence?

Excellence sets the standards of what you want and desire. Though the benefits of high expectations are essential, creating a culture of excellence is the most important. High expectations set a tone of excellence and academic rigor within the school community. It promotes a culture where students, teachers, and staff strive for continuous improvement, celebrate achievements, and support one another in reaching their highest potential.

▌ *Many teachers set high expectations for their students; parents should not water them down. Let me reiterate- outside activities should not take precedence over education. Parents are encouraged to have the same level of expectations and excellence for their child's academics as anything else. Helping your children develop high expectations at home promotes accountability, respect, and good work habits, just as teachers do in school. This joint effort helps students with goal setting, character development, and life-long dreams.*

Tip 10: Respect Has Loss Its Ranking

I remember when respecting adults was not an option, especially the teacher and all adults in the school building. If you disrespected your teachers, there were significant consequences.

There was a time when students were too afraid to talk back to their teachers, make certain gestures at them, or disrupt the lesson. But times have changed. The lack of respect that some students display towards their teachers is mindboggling.

The loss of respect in the classroom can harm the learning environment, student-teacher relationships, and overall academic outcomes. Parents and students must understand their roles, build good relationships, and communicate with teachers respectfully.

Respect must be addressed and put back at the top of the hierarchy where it belongs- always keeping its ranking intact. Parents and teachers must establish clear expectations for behavior, communication, and mutual respect from the outset of the school year.

Parents are the first teachers, and children are watching. Parents can show their children how to respect teachers through respectful communication and showing kindness. A little goes a long way. If children see their parents modeling such actions before them and demand respect towards the teachers, they will likely pass down the good things associated with developing respect.

I was walking down the hall when I heard a student telling her friends, "My mother was cussing her out and had the phone like this (improvizing) and was telling me that she (the teacher) better not say anything to me." She had the right teacher!"

Parents must model respectful behavior, active listening, and conflict-resolution skills in their teacher interactions. Let's respect its rightful place and back in its proper hierarchy.

🚪 *Parents must refrain from disrespecting teachers, mainly when their child is present. Such behavior diminishes the teacher's value and undermines their authority. It's crucial not to damage the relationship with your child's teacher. Otherwise, your child might lose valuable opportunities due to a parent's lack of awareness.*

Tip 11:

Manipulation Station

Just like children can play parents against each other, students play the same teacher against teacher and parent against teacher game.

Don't let the smooth game fool you! If a child is afraid and lacks character or integrity, they can quickly ruin a healthy parent-teacher relationship before it can develop. Children can effortlessly manipulate their parents into believing that their teachers are mistreating them. They can also manipulate parents into thinking they are doing everything they should at school, like following the rules, turning in assignments on time, and adhering to their teachers...you get the point.

I always encourage my students to tell their parents the truth because I will, and it's always better for the parents to hear the truth from them than from the teacher. Children will lie because they don't want to face the consequences.

I appreciate parents who take the time to listen to the facts and who know their children. When you know your children, you know the teacher is most likely being honest. However, there are some instances where parents will ride intensely with their children no matter what, and when the child knows that, effective communication is broken, and a wedge is formed. This broken communication or relationship can be very disheartening for the teacher who desires a partnership. It's essential to recognize that while some students may attempt to manipulate their parents about their teachers, this behavior is not healthy or productive for anyone involved.

Here are more detailed ways in which students might attempt to manipulate parents concerning their teachers:

Misrepresenting Situations: The truth may be distorted or exaggerated about their interactions with teachers or academic performance to gain sympathy or avoid consequences from their parents.

Playing on Emotions: Who's the real victim? Students may appeal to their parents' emotions by portraying themselves as victims or unfairly treated by teachers to manipulate their parents into taking their side or advocating on their behalf.

Blaming Others: Teachers are blamed for students' academic struggles or disciplinary issues by painting the teacher as the antagonist and themselves as the innocent party to avoid accountability.

Seeking Special Treatment: Students may manipulate parents into seeking special treatment or leniency from teachers, such as requesting extensions on assignments, extra credit opportunities, or exceptions to classroom rules, by playing on their parents' sympathy or willingness to intervene.

Creating Division: Sadly, students may attempt to create division or conflict between parents and teachers by misrepresenting communication or interactions between the two parties to avoid consequences or gain leverage in academic or disciplinary matters. To address potential student manipulation, parents need to recognize and address the dishonesty and manipulation for what it is.

Maintain Open Communication: Fostering open and

honest communication with your child and their teachers is necessary. Establish a supportive relationship with teachers based on mutual respect and trust, and encourage your child to do the same.

Verify Information: When your child reports concerns or issues related to their teachers, take the time to verify the information and gather perspectives from multiple sources, including the teacher and school administration, before jumping to conclusions or taking action.

Set Clear Expectations: Communicate your expectations to your child regarding academic performance, behavior, and communication with teachers. Encourage honesty, accountability, and respect for authority figures, including teachers.

Encourage Independence and Responsibility: Teach your child to own their actions, choices, and academic responsibilities. Please encourage them to communicate directly with teachers to address concerns or seek support rather than relying on manipulation or parental intervention.

Model Respectful Behavior: Model respectful and constructive communication with teachers and other authority figures in your interactions with your child. Demonstrate empathy, active listening, and problem-solving skills as you navigate challenges or conflicts together.

Seek Professional Guidance: If you suspect that your child is engaging in manipulative behavior or struggling with underlying issues such as academic pressure, anxiety, or social dynamics, consider seeking support from a school counselor, therapist, or mental health professional to address the root causes and develop effective strategies for support and intervention.

🚪 *Take time to see where you can help your child. It's always possible to show them that you and the teachers are on a team and that you value the educational partnership. In addition, there is no room for lies and manipulation. Let your child know that lies and manipulation will not be supported.*

Tip 12: I Don't Have Time

Until recently, never in my life and 20-plus years of teaching have I ever heard a student tell me, "I don't have time to do my homework." Wait, what? What do you mean you don't have time? When I heard these words, I knew priorities were not in place.

I understand that children need an outlet, and gifts should be cultivated. They are in after-school clubs and have sporting events and practices. They get home late, shower, eat dinner, and go to bed, and the process is repeated. Don't get me wrong; I'm a massive proponent of nurturing every gift a child has and allowing them to learn and experience new things. However, there must be a healthy balance, and parents, if you place these unrelated academic obligations on them, you must also teach them the balancing act early. We call this balancing act time management.

Here's a quick executive functioning introduction to time management. Mastering time management is not just a beneficial skill; it's an essential pillar of effective functioning that underpins academic success and cultivates a well-rounded life. For students, honing this skill isn't merely a suggestion but a necessity, enabling them to navigate the intricate balance between academic pursuits, extracurricular commitments, and personal well-being with finesse. By mastering time management, students can allocate their resources judiciously, prioritize tasks effectively, and seize opportunities for growth inside and outside the classroom.

Here are some immediate and practical tips you can use at home. Sit with your child and help them to:

- Set priorities
- Create a Schedule
- Break tasks into manageable pieces
- Use time blocks effectively
- Prioritize important tasks
- Avoid procrastination
- Learn to say "no"
- Stay organized
- Take breaks and rest
- Reflect and adjust

Teachers don't want to hear excuses and, to be quite honest, don't care about the things that take precedence over assigned work (except emergencies or uneventful circumstances). Teachers can't be the only ones teaching time management skills. We all have deadlines to meet and life after school. As mentioned in a previous chapter, it's a joint effort. All the tips in this list help to develop practical time management skills. Acquiring these habits will serve students well throughout their academic and professional careers. With consistency and repetition, you and your child can be productive with time management.

Tip 13:

The Parent Teacher Conference Effect

This topic was mentioned in a previous Chapter, so it must be important. Parent conferences are a great way to foster positive relationships between parents and teachers and support student success.

The truth is, there are some parents that teachers never see. I empathize with the parents of multiple children who have to juggle conferencing on one conference day and with the teacher who commits to conferencing with other parents. Still, their conferencing schedule does not permit for a conference with their own children's teachers. I've certainly been there, so no judgments here!

Parents, if you cannot attend a conference on a scheduled date, email the teacher requesting an alternative date. Teachers are willing to accommodate your request when their schedule is open.

More often than not, the parents of students performing well are the ones we see at conferences, and we appreciate you. Nonetheless, we urge parents of struggling students to make it a point to meet with us. Some parents may want to avoid hearing about the same problems shared with them year after year, such as behavior, attendance, lack of progress, or deficient skill sets. At the same time, you might gain some new insight from your child's new teachers where your child can get additional support or early intervention so they can be successful.

▌ Teachers need to conference with parents at least once during the school year. Let me reiterate- this is not a one-sided journey. You are the most critical stakeholder in this equation. Parent conferences allow you to work with the teacher to set goals, solve problems, and build trust and rapport. Therefore, take advantage of the conferences. It could be a game-changer for your child.

Tip 14: Watch Your Tone

Tone is everything and plays a significant role in communication.

The tone in communication refers to how we use our voice to convey meaning, emotions, and attitudes. Our tone can enhance or hinder the effectiveness of our message, impacting how it is received and understood by others.

Some teachers feel that a parent's tone is unwarranted and disrespectful, which can, in turn, lead to a rude response, depending upon the teacher. Teachers are human, too. I have heard on countless occasions students telling their friends, "My mother cursed Ms. or Mr. so and so out. Then, they tell friends what their parents said to the teacher. I am mindful of how I respond, but teachers must be careful about what they want to say and how they want to respond to a parent. Even when that parent has gotten on our last nerves, it's good practice not to react immediately. That delayed response, even if it's the next day or two, is what you need in that moment. Not having time to think or cool off could have a terrible outcome.

I've had parents who thought it was ok to address me in a specific tone, and I had to establish boundaries politely. One might refer to this as "checking them." For those who don't know what that means, it's not to be a doormat or to let someone know their actions or response is unacceptable. At one point, I got sick and tired of the disrespectful tones from emails. Aggressively telling the teacher, "This needs to be done now," and "You should have done this by now" is totally out of order. Excuse me! The tone! My administrators don't even address me in this tone. Remember, we talked about respect in Chapters 6 and 10. Now, let's dive back into it again because the lack of it is a significant issue. Yes, your child's teacher is a professional like you and nothing less. Please don't demoralize them.

Instead of allowing emotions to get in the way, I gather myself and my thoughts and then provided my parents with tips on teaching their children to communicate respectfully in newsletters and announcements. The tips provided for the students are also helpful for the parents. Sometimes, teachers must give parents the tools, the hints, and the models for effective communication because they might need to learn how.

For the parent:
Your tone is crucial when communicating with your child's teachers for several reasons:

Respect and Collaboration: A respectful and collaborative tone sets a positive atmosphere for communication. It fosters mutual respect between you and the teacher, promoting a partnership in your child's education.

Effective Communication: Tone can significantly influence how your message is received. A warm, open tone encourages honest and constructive dialogue, making addressing concerns or discussing your child's progress easier.

Modeling Behavior: Children learn by example and observe how adults interact with authority figures like teachers. Demonstrating a respectful and positive tone in your communications with teachers sets an excellent example for your child to engage respectfully with others.

Building Relationships: Effective communication builds and strengthens relationships between parents and teachers. A positive tone helps establish trust and rapport, which is essential for working together effectively to support your child's learning and development.

Problem Resolution: Maintaining a respectful and constructive tone can help facilitate problem resolution when discussing concerns or issues with your child's teacher. It encourages both parties to focus on finding solutions rather than placing blame or becoming defensive.

Creating a Supportive Environment: A positive tone promotes a supportive environment where everyone involved feels valued and appreciated. It encourages teachers to be more open and responsive to their concerns and requests.

For the teacher:
Your tone is crucial when communicating with parents for several reasons:

Clarity and Understanding: The tone of communication sets the emotional context for the message being conveyed. A clear, respectful tone helps parents understand the shared information and its intentions.

Building Trust: A positive and respectful tone helps build trust between educators and parents. When parents feel respected and valued in their interactions with teachers, they are more likely to trust the information provided and feel confident in the school's support for their child.

Fostering Collaboration: Effective communication with parents requires a collaborative approach. A tone that is inclusive and cooperative encourages parents to actively engage with teachers and school staff in supporting their child's learning and development.

Addressing Concerns: A supportive and empathetic tone is essential when addressing concerns or challenges related to a child's academic or behavioral progress. It helps parents feel understood and encourages them to work with educators to find solutions and support strategies.

Modeling Behavior: Just as with children, parents learn from the tone set by educators in their communications. A positive and respectful tone sets an excellent example for parents, demonstrating effective communication strategies and promoting healthy interactions.

Creating a Positive Environment: A positive and supportive tone creates a welcoming and inclusive school environment. When parents feel respected and valued in their interactions with school staff, they are more likely to participate actively in school activities and initiatives.

Teachers appreciate transparent, respectful, and proactive communication from parents. However, teachers often feel frustrated or overwhelmed if communication from parents is inconsistent, disrespectful, or accusatory. This can make it challenging for teachers to address issues effectively and create a positive student-learning environment.

Tip 15: No Exclusions on Inclusion

Students with learning differences sometimes struggle with learning and how they process information. Some come by way of ADHD, autism, and other neurodivergent diagnoses. It might surprise you, but these students are brilliant.

Neurodivergent learners often exhibit a wide range of learning styles and preferences. Some may excel in visual or hands-on learning, while others may need help with traditional classroom settings that rely heavily on auditory instruction. Learning accommodations in the form of an Individual Educational Plan (IEP) or 504 Plan are put in place so that the student has the ultimate opportunity to be successful in the classroom. The goal is to include the students in a general classroom environment with additional support.

The specific accommodations encompass a wide range of implementations that do not in any way honor or list any form of exclusion. The exclusion that I am referencing in this section is exclusion from doing the work.

Some parents feel that because their children have an IEP or 504 Plan, they are excluded from assignments, rules, expectations, behavior, etc. When students have these accommodations, parents, and students must understand that they still have a responsibility and are held to the exact expectations of their peers.

By law, the IEP and 504 accommodations must be honored and implemented in school. However, it is very challenging for a teacher when a student does absolutely nothing—not putting forth any effort whatsoever. I've been in settings where teachers have been required to change students' grades who have these accommodations, even though the students have yet to demonstrate effort, good faith, or mastery of a skill. This is upsetting for the teacher, to say the least, for many reasons. Partaking in this request gives the student and parents a false sense of gratification, which does not help the student in the long run.

There's a lot that I can elaborate on about this topic, but I'll move on for now. Nonetheless, holding these students accountable for their work is essential because it:

Promotes Independence: Holding students with IEPs accountable for their work fosters independence and self-advocacy skills. It encourages them to take ownership of their learning and responsibilities, essential skills for academic success and beyond.

Builds Confidence and Self-esteem: When students are held accountable for their work, they experience a sense of accomplishment and pride in their achievements. This can boost their confidence and self-esteem, motivating them to continue putting effort into their academic endeavors.

Prepares for Real-world Expectations: In the real world, individuals are expected to meet deadlines, fulfill responsibilities, and produce quality work. Holding students with accommodations accountable for their work helps prepare them for these expectations and ensures they develop the skills needed to succeed in adulthood.

Encourages Growth and Progress: Accountability encourages students to strive for improvement and growth. When students understand that their efforts directly impact their outcomes, they are likelier to put forth their best effort and seek support when needed to overcome challenges.

Maintains Equity and Fairness: Holding all students, including those with accommodations, to high accountability standards promotes equity and fairness in the educational system. It ensures that students receive the support they need to succeed while being held to similar expectations as their peers.

Respects Individual Potential: Holding students with accommodations accountable recognizes and respects their potential and capabilities. It acknowledges that every student can succeed and make meaningful contributions regardless of their learning differences or challenges.

Parents need to understand that students with accommodations are not excluded from performing in school. Some parents feel that since their child has an IEP or 504 Plan, it excludes them from the expectations set forth by their teachers. If your child does have accommodations, please review them with your child so they are aware of their responsibilities, identify who their support team is with them, collaborate with the teachers, and work closely with your child throughout the year.

Tip 16: You Need to Go

Most parents do a tremendous job with their children, but let's be honest. You know your child is creating terror at school- if this is you. You receive calls almost daily, and you've gotten to the point where you want to avoid hearing what they have done, so you don't answer or block the call. Meanwhile, the school has to deal with the tyrant behavior all day, which is unfair.

I'm from the old school, and if you acted up in school, you were not permitted to go on the field trip. Today, field trips are a part of the learning experience, so some school districts can no longer punish the students in this manner. As a result, they get to enjoy the fun. But at the expense of who? If your child has behavior issues, then "you need to go"! You need to go on the field trip. You need to go to field day. You need to go to the tournament. You just need to go!

I recently learned from a friend that she would block the phone number of her child's school because she got tired of receiving phone calls at work. We all laughed in disbelief, but when I thought about it, I also wondered if that is the reason why some teachers can't reach parents to address behaviors. Sometimes, emails don't get a reply, and the telephone just rings. So what's a teacher to do?

Teachers are not babysitters. They simply don't have time to call the office throughout the day or address behaviors that occur all day. We barely have time to plan lessons, attend to time-sensitive matters and deadlines. A teacher's tolerance can go so far in a given day, and their mental health is at capacity. Teachers are overworked and underpaid, so those in this profession are here as a labor of love. Some teachers' attitude is, "I need to leave!" and I'm not only referring to Millennials. A large percentage of seasoned teachers have these sentiments. When teachers are at their maximum mental capacity with student behaviors, the school system, and the lack of parental involvement, that is when we lose good teachers.

Parents, don't put your child off on the teachers when you know your child is more than a handful. Teachers should not have to shadow students with behavior issues in and outside the classroom. Just like you get a break when your child is at school, teachers need a break too! Respectfully, teachers would appreciate it if you kept your child home with you or were the first to volunteer as a chaperone. Otherwise, your child will be paired up with the teacher who needs a break or with another parent chaperone who may be unable to handle their behavior. So, in other words, "You need to go!" We'd love to have you!

Teaching children about accountability and consequences is an essential aspect of parenting. As a parent, you can help your child understand the impact of their actions and the importance of taking responsibility for their behavior. It may come in the form of being absent from an activity or having you present. We will cover the pros of consequences in Chapter 19.

Tip 17: Set a Date and Create Routines

Quality time does not always mean watching a movie or riding bikes with your child.

Quality time with your child refers to intentional, focused, and meaningful interactions that strengthen the parent-child relationship and contribute to the child's emotional, social, and cognitive development. It's about the depth and connections made in the time spent together rather than just the quantity.

Just like you carve out time to wash your child's hair or take them to practice, you should also schedule time to set routines.

Think about the skills that your child is lacking at home. Well, those same skills are also lacking at school and also need improvement, making it difficult for them to perform well. Research tells us that routines are the keys to developing skills. For example, if your child has trouble finding things and can't get out of the door on time in the morning, helping them pack their bags at night and placing them by the door is an excellent routine for developing organizational skills. Set a designated time before or after the bath. Before you know it, your children will automatically start to prepare for the following day without you having to give reminders. Keeping up with this routine will surely be a winner for you both.

I remember when my children and I struggled to get out of the door in the mornings; boy, it was stressful. I was fussing at them before the day even got started. And we all know that's no way to start the morning with your children. As I began to understand the importance of helping them set routines, it became conducive, and our mornings were smoother, calmer, and less stressful. They actually enjoyed becoming independent and showing that they were becoming responsible for prepping.

Similarly, like time management, planning, and organization are other essential executive functioning skills critical for success. Scheduling time to check grades and organize binders and backpacks is another excellent way to help your child with organization. Remember, consistency is key!

"If you fail to plan, you plan to fail." Developing these executive function routines can greatly benefit children by providing structure, stability, and a sense of security.

Here are some tips to help your child develop routines:

Start Early: Introduce routines to your child from an early age, but it's never too late to start. Let your child help with sorting, purging, and putting items back in their proper place.

Consistency: Be consistent with routines. Stick to the same schedule as much as possible, including waking up, meal times, bedtime, and other daily activities. How you do one thing is how you do everything. So, be sure to maintain the norm. It's also a good practice to organize notebooks and backpacks weekly or bi-weekly with your child until they get into the routine. Remind your child to plan and organize themselves in school by cleaning out their lockers and desks.

Create a Visual Schedule: Use visual aids like charts or posters to illustrate the daily routine. This helps younger children understand the sequence of activities and gives them a sense of control and predictability. Let them check off as they go throughout the day or task, giving them a sense of responsibility.

Involve Your Child: Involve your child in creating and adapting routines. Encourage them to participate in decision-making and give them choices within the routine, such as selecting their clothes or snacks. Consider planning out snacks and clothes with your child for the week by placing clothes in organized shelving and with snacks labeled for each day of the week.

Set Clear Expectations: Communicate expectations and boundaries related to routines. Explain why routines are important and how they help create order and predictability in daily life. If your child falters on their expectation, it is ok. Have a reminder conversation so that they are clear.

Be Flexible: While consistency is vital, be flexible and willing to adjust routines as needed. Life is eventful. Schedules change, and unexpected circumstances may require modifications to the routine. Get back to a regular routine as soon as possible.

Use Transition Strategies: Help your child transition between activities smoothly by using transition strategies such as timers, visual cues, or verbal reminders. Songs are perfect for younger children. This prepares them mentally for upcoming changes and reduces resistance and meltdowns.

Positive Reinforcement: Praise and reward your child for following routines and completing tasks independently.

Positive reinforcement encourages repetition and reinforces desirable behaviors. It also gives children the confidence to develop leadership and accountability.

Model Behavior: Lead by example and demonstrate good habits and routines yourself.
Children learn by observing and imitating their parents and caregivers, so be a positive role model. A tree is known by the fruit it bears, so be mindful of the type of seeds you are sowing in the presence of your child.

Be Patient and Supportive: Understand that establishing routines takes time and patience. Be supportive and offer encouragement, especially during transitions or when your child is learning new routines. Some parents are not patient with their children because they are not patient with themselves. Know your child and how much they can handle. Embrace their makeup and encourage them.

Parents are responsible for filling in the missing gaps. Teachers appreciate when parents go the extra mile to get their children extra support, whether tutoring or therapy. Planning and organization is also critical for student success in school and beyond. Children who learn to plan and organize are less scattered and can focus and complete tasks. They, in turn, become more productive adults.

Tip 18: Is Your Child Prepared?

This question can go in many directions, but I want to know: Is your child prepared for school?

Parents, those school supplies you bought your children at the beginning of the school year are long gone. Those supplies ran out before the Thanksgiving and Christmas holidays. Now, students are sitting in the classroom with no pencil or paper. But students come to school with new things all the time. Ouch! Teachers send sign-up sheets, newsletters, and emails requesting supplies, but they don't make it to the classroom. Teachers often purchase supplies out of their pockets for 30 or more students; not even a box of pencils comes through the door.

I'm not passing judgment, and I'm not being critical. But I am making you aware of the matter at hand. Teachers have their personal financial obligations to take care of, and it is unfair that they buy school supplies for the class, especially when they have purchased supplies for their children. You would think schools would supply teachers with basic supplies, but not all do. So what's a teacher to do? They reach out to parents, asking them to stock up or for donations. If those supplies do not show up in either form, they are the ones footing the bill so students can participate and have a conducive learning environment.

I always made it a point to donate extras to my children's teachers throughout the year on top of supplying my own children with what they needed because it is not the teachers' responsibility. Parents stay at Walmart, Target, and Amazon. Remember to visit the stationary section and grab supplies for your child while you are there.

I think the kids eat pencils! Teachers give so many pencils to the same students on a given day. Where are the pencils disappearing, too? Help a teacher out and send some supplies to school. Classrooms always need pencils, paper, hand sanitizer, markers, tissue, colored pencils, and glue. Your child's teacher will be very appreciative.

Here's something else to consider- is your child prepared for the next level? Ouch again! Is he/she on grade level? Do they have reading and writing fundamentals? How are their social and communication skills? When students lack the fundamentals of reading and writing, they struggle, and struggle can cause long-term behavioral and work habit issues. Be honest with yourself. If your child is not ready, please get them the extra support they need (tutoring, coaching, etc.). Don't let the lack of money be why your child misses out on help.

Your child's school, churches, community organizations, and centers are often good resources to tap into, and free services might be available.

Your child should always be prepared for school. This starts with a healthy breakfast, school supplies, homework, reading and math skills, and a positive mindset.

Tip 19: There Will Be Consequences

I'm not sure what school you came from, but there were consequences when I was in school. With all of the rights that children have and the lack of parental and educator rights, issuing consequences is not what it use to be.

I remember having consequences at home and school for misbehaving or violating the school rules. Yes, I was that child...well, kind of! Parents did not blame the administrators or teachers for enforcing the rules and teaching their children valuable lessons. But today, our hands are tied, and sometimes, consequences are not even rendered.

If we are setting high expectations, then there should not be a compromise. Many students want to earn points for positive behavior incentive events. Still, when they see students who could care less get rewarded along with them, the validation is discredited, and they lose momentum.

I remember a few occasions when suspended students got recognized for positive behavior. Their peers were perplexed because they knew that those students were not deserving and it wasn't fair. The message sent to the students left them with many questions. We must send the right message and implement consequences and rewards where appropriate.

For teachers to maintain a level of respect, consequences must be issued. We are preparing our future leaders for the real world, where there are consequences for arriving to work late, speeding and reckless driving, overdrawn bank accounts, and so on.

Some parents do not want their children to receive any consequences at school whatsoever. Consequences in school serve several vital purposes and can be beneficial in several ways:

Promoting Accountability: Consequences help students understand the direct link between their actions and the outcomes they experience. Holding students accountable for their behavior teaches them that their choices have positive and negative consequences.

Teaching Life Skills: Experiencing consequences in school helps students develop essential life skills such as problem-solving, decision-making, and self-regulation. They learn to assess situations, make informed choices, and manage their behavior accordingly.

Fostering Responsibility: When students face consequences for negative behavior, they learn to take ownership of their mistakes and work towards making amends or improving their behavior in the future. Consequences encourage students to take responsibility for their actions and their impact on others.

Maintaining a Safe and Positive Learning Environment: When students understand the consequences of disruptive or disrespectful behavior, they are more likely to adhere to school rules and contribute to a positive learning atmosphere. Consequences help keep order and discipline in the school environment by establishing clear expectations for behavior and enforcing boundaries.

Encouraging Respect for Rules and Authority: The

consequences underscore the importance of following rules and respecting authority figures such as teachers, administrators, and school staff. When in place, students learn that rules are in place to ensure fairness, safety, and respect for everyone in the school community.

Supporting Social and Emotional Development: Through experiencing consequences, students develop greater self-awareness and emotional intelligence. Consequences provide valuable opportunities for students to learn from their mistakes, develop empathy for others, and understand the impact of their actions on themselves and those around them.

Promoting Positive Behavior: Consequences can be used to reinforce positive behavior and encourage students to make constructive choices. Recognizing and rewarding positive actions motivates students to exhibit desirable behaviors and contribute positively to their school community.

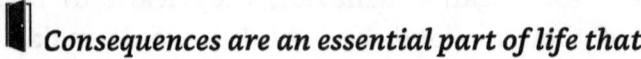 *Consequences are an essential part of life that*

Tip 20: No Bullying Allowed

Parents, I understand you want what you want for your kids and the best for them.

Generally speaking, parents want their children to make all A's, win the championship game, or be the star player on the team. You want them to have the opportunities that you never had or better, and you want to brag about their successes. Let me sound the alarm- please stop trying to strong-arm the teacher when the cards don't line up.

Parents may attempt to bully teachers in various ways, though it's essential to note that this behavior is not representative of all parents and is unacceptable. Some ways in which parents may try to bully teachers include:

Aggressive Communication: This may involve yelling, threatening language, or being excessively confrontational during meetings or interactions with the teacher.

Intimidation Tactics: Some parents may attempt to intimidate teachers by leveraging their position or influence within the school community, threatening to involve higher authorities or take legal action.

Blame and Accusations: Parents may unfairly blame teachers for their child's academic struggles or behavioral issues, making unfounded accusations or criticisms.

Demanding Special Treatment: Some parents may insist on preferential treatment for their child, such as requesting higher grades or special accommodations, and become hostile if their demands are unmet.

Spreading Rumors or Gossip: Sometimes, parents may spread rumors or gossip about the teacher, attempting to damage their reputation or credibility within the school community.

Cyberbullying: With the rise of social media, some parents may resort to cyberbullying tactics, such as posting negative comments or reviews about the teacher online or through email.

Undermining Authority: Parents may challenge the teacher's authority in the classroom, disregarding disciplinary measures or undermining instructional decisions.

Harassment and Personal Attacks: In extreme cases, parents may engage in harassment or personal attacks against the teacher, targeting them with insults, derogatory remarks, or even threats of physical harm.

These behaviors are unethical, and teachers should not be subjected to them in any way. We all have a common goal for our students- success. So be delightful, not a bully!

Final Thoughts:

Dear Parents,

I hope this book has served as both an encouragement and an eye-opener for you. Your commitment to learning and growth for your child is invaluable. I am genuinely grateful for your dedicated time reading and absorbing the helpful tips I've shared.

Let this encourage you! Implement the tips provided with confidence and enthusiasm. The benefits you and your children will experience by putting these strategies into action are immense. Every small step you take towards fostering a nurturing environment for your child's growth is a step towards a brighter future.

Remember, you are not alone in this journey. There are numerous resources available to support you as you apply the advice from this book. You can find additional guidance and connect with me through my website, www.wendygrayconsulting.com, and social media pages as "Your GoTo Educator".

Your dedication to your child's development is commendable, and I want to emphasize the positive impact you can have by applying the knowledge you've gained from this book. Your efforts matter, and they make a real difference.

Lastly, I encourage you to stay connected and share your experiences with me. Your feedback is invaluable, and I am here to support you every step of the way. Please don't hesitate to reach out if you have any questions or need further guidance.

Again, thank you for your unwavering commitment to your child's well-being and growth.

Warm regards,

Wendy

www.ingramcontent.com/pod-product-compliance
Lightning Source LLC
Chambersburg PA
CBHW070417230526
45471CB00006B/2855